Fire in the Trees

MARY JO NICKUM

Aquitaine Ltd
Phoenix, Arizona

Fire in the Trees

MARY JO NICKUM

Aquitaine Ltd
Phoenix, Arizona

Cover design by JD Smith Design

Library of Congress Control Number: 2017942670

ISBN-13: 978-0-9980858-2-1
ISBN-10: 0-9980858-2-0

www.aquitaineltd.com

DEDICATION

This book is dedicated to young people everywhere who
Don't like to or find it difficult to read and
Who, therefore, live on the fringes of a happy, healthy life.
Learn to enjoy reading and the world can be yours.

Believe in yourselves

Why a Reluctant Reader Series?

Not all children read at the same level nor do all children enjoy reading. While a good many children do become excited and engaged in reading, especially in the primary grades, some are reluctant and disinterested. While a child may not show a natural interest in reading, this does not mean he/she cannot become a skilled and even enthusiastic reader. If the child has reached middle grade and is still disinterested, it's time to take action.

While any child, young or old, male or female can be a reluctant reader the largest number of unenthusiastic readers are adolescent boys. Research shows that a good number of boys who were avid readers in the elementary grades become disinterested in reading during their middle school years. However, there are a number of factors that may contribute to this shift—increasing complexity of material, peer pressure—one of the primary reasons seems to be they fail to see the connection between reading and "real" life.

School assignments, such as book reports, can become stressful for these students as well as parents and teachers. These children need material that is especially prepared to be relevant to the curriculum as well as to life. This material must be written to engage the reluctant reader, using images as well as text and presented in electronic and paper to fit classroom and leisure reading.

Books for the reluctant reader must be:

- Relevant to the curriculum (Common Core),
- Factual but engaging,
- Written to pique and hold interest,
- Presented in text and images,
- Available as ebook and in print, and
- Priced to be affordable for the individual student as well as bulk-priced to be attractive to schools.

Aquitaine, Ltd.'s Reluctant Reader Series, geared to middle grade readers, fills all these requirements. Our books have been thoroughly researched and edited by leading scientists and written by educators and librarians.

Illustration List

Figure 1a After the fire 3
Figure 1b After the fire 4
Figure 2 Tree Trunk 5
Figure 3 Needles 6
Figure 4 Closed cone 6
Figure 5 Cone on needle bed 7
Figure 6 Open pinecone 7
Figure 7 Pine pollen 8
Figure 8 Male cones 9
Figure 9 Chipmunk 10
Figure 10 Woodpecker 11
Figure 11 Seeds in pine needles 12
Figure 12 Mule deer with fawns 13
Figure 13 Deer-tracks 14
Figure 14 Birds searching for seeds 18
Figure 15 Snow in the forest 19
Figure 16 Seeds germinating 20
Figure 17 Seed root curving 21
Figure 18 Pine seed above ground 22
Figure 19 Pine needles on seedling 23
Figure 20 Several pine seedlings 24
Figure 21 Dense pine forest with nothing growing underneath 25
Figure 22 Seedling becomes a tree 27

Figure 23 Pine sapling	28
Figure 24 Tree development	29
Figure 25 Tree structure	30
Figure 26 Logger's belt	33
Figure 27 Pile of logs	34
Figure 28 Helicopter logging	35
Figure 29 Sawmill	36
Figure 30 Making furniture	37
Figure 31 Paper we use every day	38
Figure 32 Dens for wildlife	38
Figure 33 Birds nesting in snag	39
Figure 34 Clear cutting	40
Figure 35 Moutain pine beetle	41
Figure 36 Dead trees killed by mountain pine beetle	42
Figure 38 Pine cone with seeds	43
Figure 39 Windfalls - fallen trees	44
Figure 40 Photosynthesis	45
Figure 41a Forester at work	56
Figure 41b Forester at work	57

Pine on the Mountain

Table of Contents

Prologue	1
Chapter 1 Recovery	3
Chapter 2 Opening	17
Chapter 3 Growth	20
Chapter 4 Passage of Time	26
Chapter 5 More Time	32
Chapter 6 Logging	36
Chapter 7 After Logging	43
Chapter 8 The Future of Trees	46
Chapter 9 Becoming a Tree Scientist	55
Glossary	63
Sources	66
Illustration Credits	69
About the Author	73

Mountainside Forest

Prologue

Dark clouds advance—lightning flashes, brightening the sky. A westerly wind whips the tree branches. There's a brief lull, then a sudden streak of lightning jabs the top of one of the trees. With a gust of wind, several hot branches fall to the ground. The top of the tree and the branches on the ground smolder. More bright red embers drop from the tree to a forest floor littered with last year's brown, dry needles. Some will cool to become dry ash, others will not. The wind fans the embers. The embers heat the dry needles. A flame flickers, goes out, flickers again, this time stronger. Then more needles begin to burn.

The fire gets bigger and, pushed by the wind, begins to creep further among the trees. Most of the trees on the ridge are young and strong, but a few are decaying after having been killed by a bark beetle infestation several years ago. They're just tall, dead snags now. The fire lurks on the ground while the storm subsides and night envelops the ridge. Humidity rises enough during the night to slow the advancing fire.

As the morning sun peaks over the ridge, all is quiet. No smoke rises above the treetops. A ranger in a forest surveillance plane overhead sees no problems and radios the information to the fire dispatch center. The fire continues to smolder undetected. As the day warms, a breeze flutters through the trees. It gets stronger quickly, reaching the smoldering needles. The fire gains new life. It continues its slow crawl toward the dead tree called a snag. As the daytime temperature continues to rise, the fire gets stronger and moves faster. By late afternoon, the fire reaches the dead bark around the base of the snag. Now though, the breeze has gone down and the sun is setting. The fire slows.

The next day is windy. Nothing stops the fire now. The dry bark makes fine tinder. The fire blossoms into tall orange flames as it licks the base of the snag. The snag becomes a tall column of flame. The wind carries heat and embers on to other dead or dying trees while the fire continues its path through the forest. As the heat from the burning resin in the trees increases, more and younger trees fall victim to the fire. Now a large plume of smoke can be seen rising from the ridge. A forest fire is born.

Chapter 1

Recovery

It has been several years since the fire. New trees are growing and the forest is green again. Only a few charred stumps and snags (Figure 1a and 1b) can be seen. The fire didn't cause soil damage as sometimes happens. Snow melt and heavy Summer rains can cause soil erosion after a fire. The heavy forested slopes below the fire kept the soil from washing away this time.

Figure 1a After the fire

Figure 1b After the fire

What is a tree? Trees are the tallest of the plant life on the earth. A tree is a big plant with a stick in the middle. This stick is called the trunk (Figure 2). What do we mean by "big"? By putting a figure on it, we say a minimum height of 16 to 19 feet.

Figure 2 Tree Trunk

How do trees grow? The life story of the tree begins like this. On an August morning, a light breeze sings in the pines. It foretells another hot day in the forest. As the air heats, the breeze grows stronger. A few needles drop (Figure 3). A gust of wind shakes the branches and a cone falls to the ground. Suddenly, a strong gust howls through the trees and many cones fall.

Figure 3 Needles

The seeds of the pine tree are enclosed in the scales of the cones (Figure 4) that lie on a bed of dry, fallen needles (Figure 5). The cone had hung from the tree for several years until the wind finally shook it loose. Future pine trees will grow from some of the seeds. First, though, the seeds have to be released from the cone. This, too, takes time, maybe months or even years. As the cone begins to dry, the scales crack open (Figure 6).

Figure 4 Closed cone

Figure 5 Cone on needle bed

Figure 6 Open pinecone

Pinecones are the mature female reproductive organs of the tree. The flower-like male conelets produce an abundance of pollen, which covers the ground and any surface, with yellow dust (Figure 7). The female cones capture this windblown pollen and enclose the germinated seeds tightly, sometimes for years, until they ripen. Male cones (Figure 8) are small and drop off quickly after shedding their pollen

Figure 7 Pine pollen

Figure 8 Male cones

A chipmunk (Figure 9) scampers under the pines. It stops to pick at a few of the cones. The chipmunk quickly loses interest in them and dashes off. A bit later, a small bird flits through the trees and lands briefly on a cone. It pecks at the opening scales. The bird nearly gets to the seed when the arrival of a rabbit startles it into flight. The cone continues to lie on the ground, drying in the Summer heat.

Figure 9 Chipmunk

Time passes, the bird flies south for the Winter. The chipmunk prepares a nest under a decaying stump. The rabbit's coat grows thicker in anticipation of the upcoming cold. Snow begins to fall. The wind is no longer warm, but slices through the forest like sharp knives. As more snow falls, the forest is quiet. On sunny days, the chipmunks chatter and frisk about on top of the snow. A woodpecker (Figure 10) knocks on a tree, trying to find insects for lunch. As the short days lengthen into night, an owl hoots and a coyote barks at the rising moon. Time passes toward Spring and snowmelt.

Figure 10 Woodpecker

Spring comes late to the mountains. In some places, especially in the shade, the snow doesn't melt until late June. Yet when Spring does arrive, the forest comes to life quickly. Birds build nests, animals bear their young and the ground dries. The cones from last Summer remain on the ground untouched. Birds and small animals take a renewed interest in them. The cones are turned over and over, gnawed, scratched and pecked. As the Summer heat returns, the cones finally ripen. The scales open like a set of venetian blinds and the seeds drop on the layer of needles below.

The seeds lie in the needles (Figure 11). If you could see them, they would look dead. They don't seem to move or grow. Summer rains come, gentle at first, but fall harder as the storms gain force. The raindrops hit the seeds repeatedly. Each drop forces the seeds deeper into the needle layer. After several rains, the seeds are lying on the soil under the needles.

Figure 11 Seeds in pine needles

In the Summer, larger animals roam through the forest from time to time. Today, a mule deer with her growing fawns (Figure 12) meanders beneath the pines. The fawns begin to play, jumping and prancing. As they frolic, their small pointed-toed hoofs puncture the soft forest soil. Unknowingly, they plant pine trees with their hoofs. The seeds are now in the soil (Figure 13) and ready to sprout. However, not all of the seeds will produce seedlings (Side Bar 1).

Figure 12 Mule deer with fawns

Figure 13 Deer-tracks

Side Bar 1

Forest Facts

- A bolt of lightning can be as much as 53,540o Fahrenheit.
- Deciduous forests have trees that lose their leaves in the Fall. These trees have four layers – very tall trees like oaks, smaller trees and saplings, shrubs and bushy plants, herbs and small ground plants and finally the ground layer, which includes mosses and lichens.
- Many plants and animals live in forests. Many animals hibernate in the Winter.
- Some forests are plantation forests. These are forests that have been planted by tree farmers. The plantations supply much of the wood humans use.
- Some forests have evergreen trees, such as pine trees. These trees don't lose their leaves, or needles, in the Winter.
- Redwood trees growing in the Redwood forest in northern California can grow more than 350 feet tall. Average skyscrapers are about this tall. Some of these trees are 2,000 years old.
- **More than 25% of the medicines we use originate in rainforest plants.**
- **The Boreal forest (aka Taiga) is the largest land habitat on the planet.**
- **More species of plant and animal live in the rainforest than any other land habitat.**
- **Not all rainforests are tropical.**
- **The forests are the lungs of our planet.**
- **The tallest tree in the world is called Hyperion,** a coast redwood from California that measures an incredible 379.29 feet tall.
- **Pando, also known as The Trembling Giant, is a colony of quaking aspen trees.**
- Forests cover a third of the land our planet.
- About 1% of forests (an area roughly the size of New Zealand) are consumed by fire each year.

After the Fire

Chapter 2

Opening

The soil is warm when the seeds are pushed into the ground. Nevertheless, Summer is over and Fall begins. The days are cool and nights are cold. The seeds can't grow now. They will be inactive for the time being. They need warm soil and water to sprout. Another Winter passes. Snowfall is light and Spring comes early. The days warm more rapidly than usual. Very little Spring rain falls and no rain reaches the ground below the branches of the big tree.

Chipmunks and squirrels scurry about looking for food. Birds come to search for seeds that have now all fallen from the cones. Some of the birds scratch, overturning the needles in search of more seeds (Figure 14).

Figure 14 Birds searching for seeds

The seeds absorbed a little moisture. Still there is not enough moisture and the seed covering remains unopened. The humidity is low and the top layer of needles become brittle. However, the ground below the blanket of needles does not dry out completely. The seeds don't grow but don't dry up either. They stay undisturbed in the ground as the dry, hot Summer ends. With the Fall, some rain comes but not enough to help the seeds.

Now the snow falls (Figure 15) early in the Winter. It is frigid yet the seeds are protected by a thick covering of dry needles and a little soil. The snow is deep this Winter, but Spring comes early and suddenly. The snow melts rapidly, running in streamlets down the hills. Very little water soaks into the ground. One seed receives some moisture from the melting snow and starts to grow.

Figure 15 Snow in the forest

The moist forest soil warms under the trees and needles. A seed grows rapidly at first, putting out the first tiny root. It's still not able to absorb food from the soil with this tiny root. It continues to use the nutrients stored inside the seed. Still, to continue to grow, it must soon be able to get food from the soil or it will starve. The Summer comes, hot and dry. The seed stops growing and the little root dries up. The Fall showers and Winter snows come again and another year passes. Will any of the pine seeds grow this year?

Chapter 3

Growth

It is now five years since the cone fell from the tree. Spring comes again. As the snow melts, the water sinks deep into the soil. After the snow is gone, the Spring sun warms the forest soil. More seeds begin to grow. A tiny root grows out of a seed (Figure 16). It grows longer and longer. Then another one bursts out of the seed covering. Now both roots bring food back to the seed. Other things are happening inside the seed too.

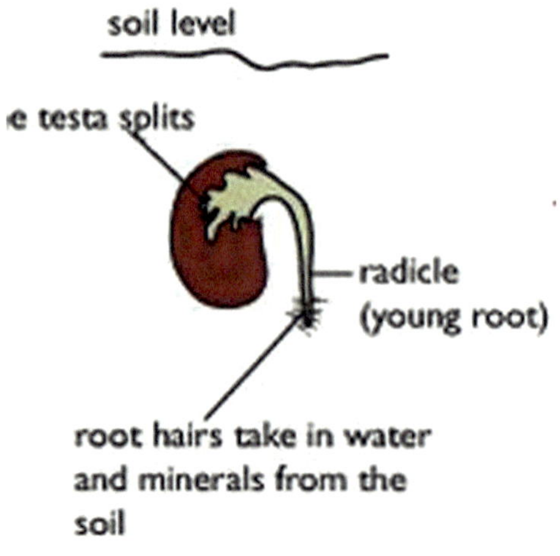

Figure 16 Seeds germinating

One tiny green shoot grows from the seed. It curls around the seed coat until the shoot nearly reaches the root (Figure 17). The shoot grows stronger with the food carried by the roots from the forest soil. It begins to straighten and push upward through the thin layer of soil (Figure 18). As it breaks through the soil, it receives some daylight and it starts to unfold. Two or three green needles (Figure 19) appear on top of the shoot. It pushes past the layer of fallen needles that helped shelter it as a seed. Now it needs sunlight so it can use the water and nutrients collected by the roots. The nutrients in the seed have been used completely, long ago. The shoot is now a seedling and begins its journey to becoming a tree in the forest. Other seeds have completed the same process. There are now several seedlings in a small area among the larger trees (Figure 20).

Figure 17 Seed root curving

Figure 18 Pine seed above ground

Figure 19 Pine needles on seedling

Figure 20 Several pine seedlings

Light is necessary for the seedlings to continue growing. The soil remains damp under the larger trees long after the snow has melted. Moisture and food from the decaying needles and twigs is available to all of the seedlings. Light, however, can be a problem. The large trees provide so much shade, causing some of the seedlings to die. The seedlings that grow further from the tree and can get light for several hours each day, continue to grow. That's why in a forest, almost nothing grows directly under large, tall trees (Figure 21). Even grass will die for lack of light.

Figure 21 Dense pine forest with nothing growing underneath

Now the seedlings must survive their first Winter above ground. The snow protects the tiny trees from the freezing winds. There is danger the heavy snow will bend or break the seedlings. Some do break. They will not recover when the snow melts; they will not become forest trees. Other seedlings will bend and, after the snow melts, be unable to stand straight. They will be stepped on and squashed by larger animals walking through the forest. Only a few of the seedlings survive the snow cover because their stems grew the strongest last year. Now that Spring has arrived, they continue their quest for food and light.

Chapter 4

Passage of Time

Several years have passed. The seedlings have become small trees (Figure 22), called saplings. They have several branches and a tip made up of several needles, held close together by a thin skin-like material. These will burst out to form new branches as the tree continues to grow taller. The sapling has a slender trunk and several branches (Figure 23). While it's slowly putting out new branches, many things are going on inside the tree. The sapling is not only growing taller, it's also getting fatter.

Figure 22 Seedling becomes a tree

Figure 23 Pine sapling

Trees don't grow like we do. As we mature, our bones, skin and muscles all increase in size. Trees grow by producing new cells in only a few places (Figure 24). These places of cell division are called meristems. Meristems are zones of intense activity. They are where all new cells are formed and where they expand (Side Bar 2).

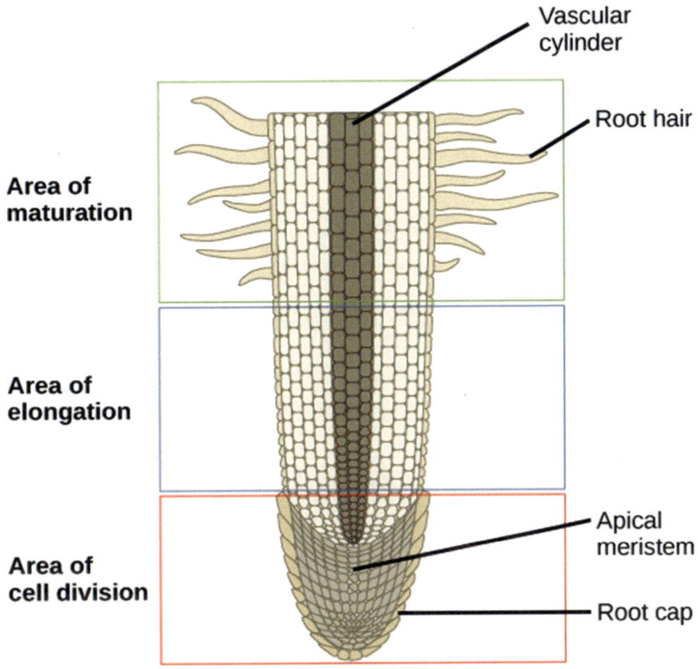

Figure 24 Tree development

Side Bar 2

Meristems - Where Cells for New Organs Originate

- Regions where cell divisions produce plant growth
- Apical meristems
  - Lengthen stems and roots
  - Repsonsible for *primary* growth
- Lateral meristems
  - Increase width of stems
  - Responsible for *secondary* growth

Trees grow in height as a result of meristems located at their branch tips. These meristems are called apical meristems. Roots also expand through the soil by growing at their tips because of the apical meristems. All buds you see on a tree contain apical meristems.

Trunk diameter growth occurs because of another kind of meristem called the vascular cambium. The vascular cambium produces new xylem and phloem each year and, as a result, the trunk, branches and roots continue to increase in diameter (Figure 25).

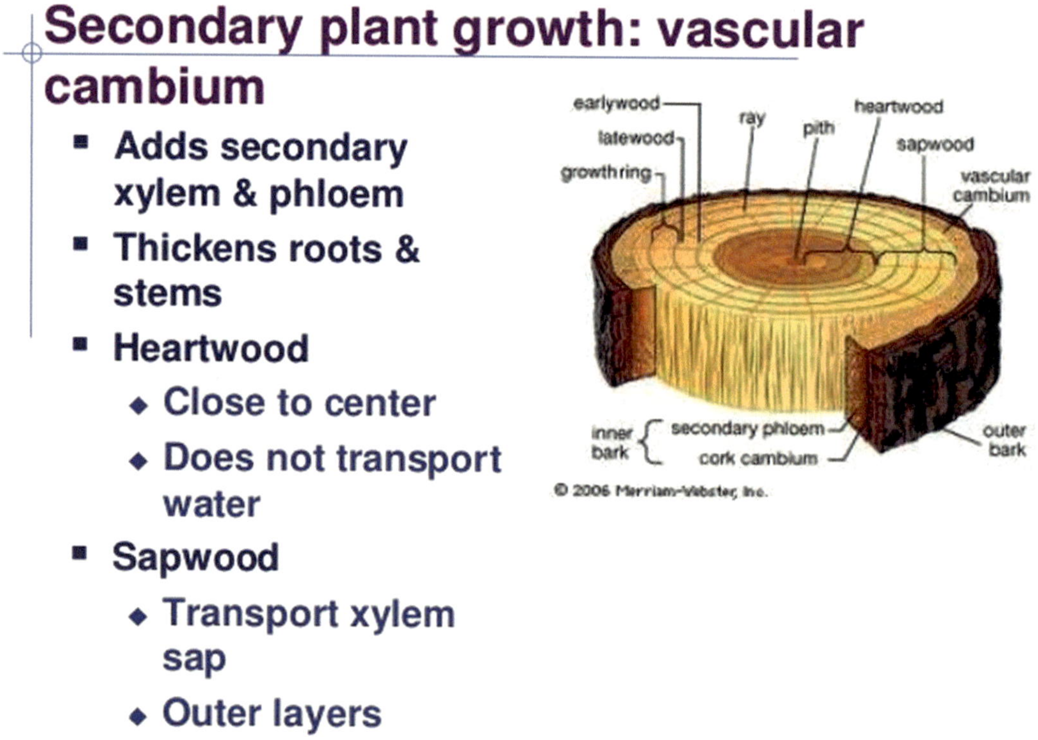

Figure 25 Tree structure

The buds grow out, yet the whole twig doesn't extend. This is the reason why a nail pounded into a tree trunk never changes height. New growth is added to the tips; the other way a tree grows is in diameter. Usually, you can't see it, but as the twig elongates it is also getting thicker. Diameter growth continues for as long as the leaves are on the tree. When the weather becomes unfavorable for growth (dry or cold), elongation slows down and a new bud starts to form. The new bud will grow again when Spring comes and the weather warms.

Chapter 5

More Time

Many years have passed. Two saplings have grown to be young trees, over 10 feet tall with a thick trunk, 4 inches in diameter. Now, they are considered trees. Of all of the seeds that fell from the cones, only two have survived. These trees grew where conditions were best for them. They had plenty of moisture, light, nutrients and protection from the harshest elements of heavy snow, hard rain and hail. Their stems grew to be solid trunks, protecting them from bending and being damaged by animals.

One day, two men came to the forest. They wore yellow hard-hats, gray sweatshirts with writing on their backs and leather gloves. One carried a clipboard; the other wore a large leather belt with several tools hanging from it (Figure 26). Both men looked at the tallest trees. One wrote on his clipboard as the other one used some of his tools. They were studying each tree on the ridge, noting each and marking many of the larger ones with blue paint. The men left the forest.

Figure 26 Logger's belt

The days grew shorter and cold winds began to blow. The temperature reached 26 degrees before it snowed. Finally, as the snow fell, a bulldozer roared and crashed through the forest. It stopped before it reached the two young pines. Chipmunks, squirrels and rabbits ran from the loud noise, birds squawked and flew a short distance before stopping to watch. Several workers, all dressed like the first two men, started their power saws. They were loggers. The saws roared and whined as trees crashed to the ground. Large tractors pulled the fallen trees to a flat area created by the bulldozer. The fallen trees, now called logs, were piled four or five logs high (Figure 27).

Figure 27 Pile of logs

After two weeks of whining saws, crashing trees and roaring tractors, the forest was quiet again. Animals returned to search for food. Some were forced to make new Winter shelters because their old ones were damaged or destroyed by the tractors dragging logs to the piles.

One day, several men with yellow hard-hats came to the pile of logs. Suddenly, a loud roar came over the forest, above the trees. It stopped over the trees near the pile. A long cable was lowered from the helicopter (Figure 28). The men on the ground wrapped the cable around the pile. One man waved his hand and the pile began to rise. The noise grew louder as the helicopter raised the bundle of logs above the forest. As the helicopter flew away, the noise got further away. The loggers left.

Figure 28 Helicopter logging

Chapter 6

Logging

Where did the helicopter take the logs? The logs were airlifted off the mountain because no roads could be built without destroying more of the forest. The helicopters took the logs down to a lower, valley area where they could be loaded onto trucks and hauled to a sawmill (Figure 29).

Figure 29 Sawmill

Logs have many uses. We use trees for lumber to build houses and make our furniture (Figure 30). Smaller logs, called pulp, are used to make paper for books, writing pads and various things we use every day (Figure 31). There are many uses for wood. It is fun to use small sticks to make a campfire. Dead trees, stumps and piles of branches are used often by wildlife for dens (Figure 32). Many birds build nests in the branches of taller trees or snags (Figure 33), where they feel safe.

Figure 30 Making furniture

Figure 31 Paper we use every day

Figure 32 Dens for wildlife

Figure 33 Birds nesting in snag

The biggest and tallest of some of the trees were removed by the loggers. This is called "selective cutting" as opposed to "clear cutting", which is a logging practice where all trees are removed (Figure 34). Large openings above allowed more light to flood the forest floor. The young trees could grow faster. There was more sunlight and more room to grow. The young trees now could stand straight and aim for the sky. Nothing could stop them or harm them. Or could it?

Figure 34 Clear cutting

One hot Summer day, a small, black mountain pine beetle (Figure 35) begins to attack a few pines on the lower part of the trunks. It had been dry and all of the trees needed water. No rains came, only the beetles. More beetles came, climbing on almost all of the trees. The beetles bored into the bark of some of the trees. The trees were weakened by lack of water and nutrients. A few had been injured during the logging operation, the Summer before. Injuries included scrapped bark, which gave the beetles easy entry into the trees. The needles of some of the trees turned yellow then brown and fell to the ground; no green needles replaced the brown ones. Those trees died (Figure 36).

Figure 35 Moutain pine beetle

Figure 36 Dead trees killed by mountain pine beetle

The beetles couldn't bore into the bark of the young trees. Their bark was strong and thick. The young trees were growing well in their forest opening. They were the new generation of trees to blanket the mountainside. A forest service plane flew overhead and a fine spray was released on the trees (Figure 37). Dead beetles littered the ground. But many were still under the bark of some of the pines. They would live there until next Spring when they would hatch and a new, smaller crop of beetles would emerge. It would take several years and more spraying to kill the beetle infestation completely.

Chapter 7

After Logging

The trees on the ridge are tall and healthy young trees. They have recovered from the fire, the logging operation close by and the beetle infestation. They have produced their first cones. The cones are small and empty. As the young trees grow older though, they will produce cones with seeds (Figure 38). The cones will open and the forest will continue.

Figure 38 Pine cone with seeds

Trees in the forest sometimes suffer injuries from diseases and insects, fire, wind and logging. Even if some of the trees are not cut down, they may be trampled or harmed by the logging equipment. Yet these ravages usually don't do permanent harm. The trees come back slowly. When most of the trees are logged or a fire has burned all of the trees in a large area, foresters will replant trees in the damaged area. This helps the forest to return more quickly. It also keeps erosion from causing further damage.

Severe storms with strong, damaging winds can topple large trees. It is unusual though, for a storm to cause extensive damage to large areas of a forest. Tornados and hurricanes do cause this kind of destruction. Wind from other storms can cause branches to break or a few trees to fall. Wind damage in a forest can be seen for years after the storm, but the forest recovers quickly because the fallen trees (Figure 39), "wind falls", have made way for new, young trees to take their place.

Figure 39 Windfalls - fallen trees

The forest is important. It provides many things, from the wood we need for furniture and houses to the fresh air we breathe. Yes, the forest helps clean the air. You see, when trees grow, a process scientists call photosynthesis (Figure 40) takes place. This process takes place within the leaves or needles. Trees take in carbon dioxide and hydrogen from the air and mix it with water and chemicals from their roots. This is how trees make their food. While this process is taking place, trees give off oxygen. That's how they give us fresh air.

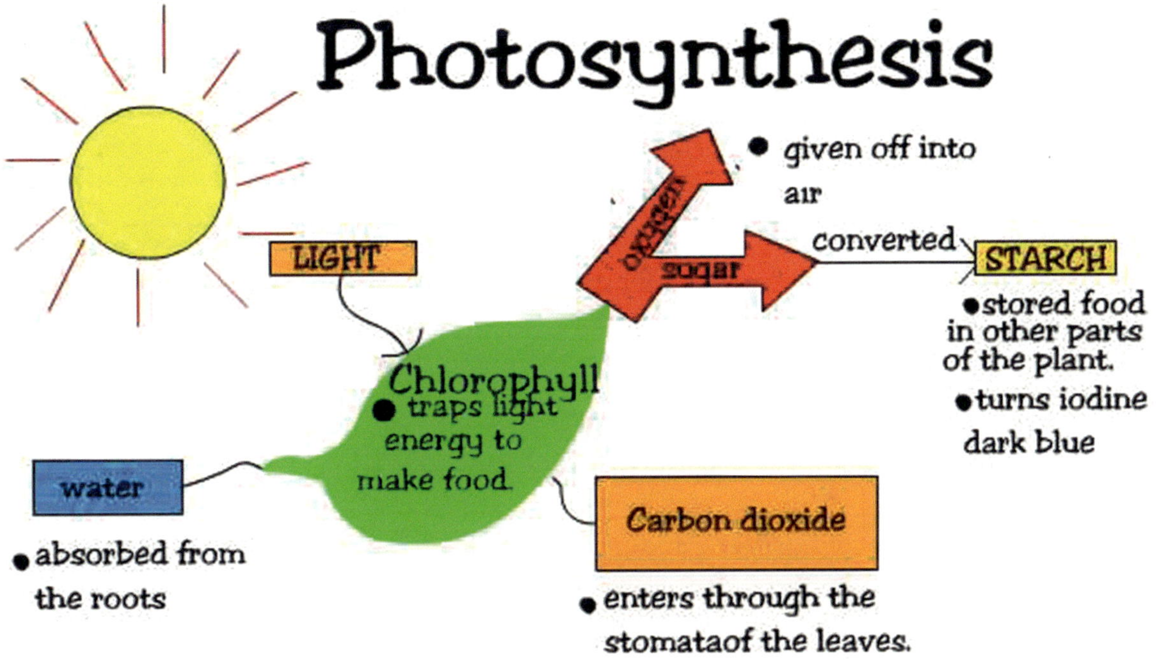

Figure 40 Photosynthesis

In addition to wood for building, trees provide a variety of materials that we use every day. Not all trees provide the same products. We use wood from some trees to make the paper used for books and writing. Other trees provide fiber, cork, dyes, resins and varnish, medicines, oils, and preservatives. Trees give us shade and several give fruits and nuts; and, best of all, our traditional Christmas tree.

Chapter 8

The Future of Trees

Around the world, trees are under threat in the wild and in urban areas. Only concerted effort can preserve tree diversity, urban forests, and the many benefits they provide to people and the environment. But those efforts are hampered by a lack of scientific understanding of the complex challenges faced by trees and forests and by a diminishing pool of people trained to meet those challenges.

The urban forest improves the lives of more than 220 million people in the United States alone, or 80 percent of the population, and with urban populations projected to grow, the number of people who benefit from trees in urban areas will increase (Side Bar 3).

Trees play important roles in the urban ecosystem because they:

- Clean pollutants from the air,
- Reduce energy use and noise pollution,
- Improve water quality and help control flooding,
- Increase property values; provide wildlife habitat,

- Mitigate changes in climate and,
- Improve community health and well-being.

Side Bar 3

Fifty Ways Trees Benefit Our Health, Community and Environment

With about three trillion trees on the planet, fortunately most of us don't have to go far to find one.

However, even though they exist on almost all corners of the earth, trees are still heavily exploited and at risk from human activity. Humans cut down about 15 billion trees every year and, according to researchers, before we came around, there were about twice as many trees on the planet.

Thankfully, researchers are continuing to prove trees are crucial for our planet's survival in an effort to stop deforestation. If that is not reason enough to save them, they may also affect everything from our stress levels to our air conditioning bill.

Personal Benefits

1. Hospital patients recover more quickly and with fewer complications, if they can view trees outside.
2. Trees in the landscape relax us; they lower our heart rate and reduce stress.
3. According to one study conducted in London, people who live in areas with more trees have fewer antidepressant prescriptions.
4. Trees, such as pine, cedar and Eucalyptus , provide beautiful aromas to soothe and delight us.
5. Trees also supply inspiration for the creative arts such as painting and ceramics.

6. Children with ADHD show fewer symptoms when they have access to nature.

7. Your favorite tree can give your kids a chance to make bark rubbings or leaf rubbings, using crayons and paper.

8. Whether as houses for children or creative and spiritual inspiration for adults, trees provide a space for human retreat.

9. When you consider the visual impact of trees and how long they last, trees are one of the least expensive plants to add to your yard.

10. Trees provide food: An apple tree can yield up to 15-20 bushels of fruit per year. Aside from fruit for humans, trees provide food for birds and wildlife.

11. A tree can add music to your life by attracting singing birds.

12. Trees can separate and define space, which can give you a sense of privacy, and create a feeling of relaxation and wellbeing.

13. Trees are beautiful to look at (most of the time).

14. Climbing trees provides wonderful, fun exercise.

15. Some trees are considered sacred. The specific tree varies between cultures, but those believed to be sacred may be especially beautiful, have an unusual size, or other distinct characteristics.

16. Studies have found that workers are more productive when they can see trees or plants from their office windows.

Environmental Benefits

17. Trees are carbon sinks, meaning that they absorb more carbon than they release.

18. Trees clean the air by absorbing odors and potentially harmful gases (nitrogen oxides, ammonia, sulfur dioxide and ozone).

19. Trees fight climate change: Excess carbon dioxide (CO_2) is building up in the atmosphere, contributing to climate change. Trees absorb CO_2 and release oxygen back into the air.

20. Trees can also filter particulates (dirt) out of the air by trapping them on their leaves and bark.

21. Deciduous trees act as a natural air conditioner by blocking sunlight in the Summer and allowing sunlight to reach and warm your home in the Winter.

22. According to the U.S. Department of Agriculture, the net cooling effect of a young, healthy tree is equivalent to ten room-size air conditioners operating 20 hours a day.

23. By reducing heating and cooling costs, trees can reduce dependence on oil and natural gas.

24. Cooling also occurs when water evaporates from the leaf surface.

25. Fallen tree leaves can reduce soil temperature and soil moisture loss. Decaying leaves promote soil microorganism.

26. At the same time, they increase ground water recharge and cut back on the number of harmful chemicals transported to our streams.

27. The presence of trees gives shade, which creates microclimates where you can grow shade-loving plants.

28. Trees help reduce surface water runoff from storms, thus decreasing soil erosion and the accumulation of sediments in streams.

29. Trees provide habitat for wildlife: Sycamore and oak are among the many species that provide excellent homes for birds, bees and squirrels.

30. Other animals that live in trees include the flying lemur, the koala, the spider monkey, and the tree kangaroo.

31. Even snags, or standing dead trees, provide a home for several different

species. Beetles and tree frogs can live under a snag's bark, while woodpeckers and other birds feed on the insects that live in snags.

32. Park hour for squirrels: Watch the squirrels in your yard or at a park skitter along tree branches at amazing speeds, on those "squirrel highways".
33. The lives of trees are bound intimately with the growth of fungi. Trees and fungi exchange vital nutrients between their roots, and each depends on the other.
34. If you love to eat truffles, that delicious French delicacy, you may be aware they cannot be planted; instead, farmers plant oak trees, because truffles only grow near the roots of certain oaks.
35. Trees can provide a windbreak: the more compact and dense the foliage on the tree or group of trees, the more effective the windbreak.
36. All forms of precipitation, rain, sleet, hail, are slowed by trees, providing some protection for people, pets, and buildings.
37. If the U.S. had 100 million trees living around city spaces, we would save about $2 billion in energy costs each year.

Community Benefits

38. The strong tie between people and trees is often evident when community residents speak out against the removal of trees or rally to save a particularly large or historic tree.
39. Trees make communities more livable; People walk and jog more on streets with trees.
40. Trees bring natural elements and wildlife habitats into urban surroundings, increasing the quality of life for residents.
41. Trees increase the value of property. Houses surrounded by mature trees may sell for 7 -19 percent higher than houses with no trees.

42. Children and adults have a cool place to play or relax in the Summer and, too, this may increase their interaction with neighbors.

43. Trees can camouflage ugliness like concrete walls, parking lots and unsightly views.

44. They also muffle sound from nearby streets and freeways, creating a comfortable canopy of green.

45. Trees cool city streets.

46. Trees provide unity. As landmarks, trees can give a neighborhood a new identity and encourage civic pride.

47. Trees can create a lasting positive impression on how a community is perceived by visitors.

48. A community's economic stability is enhanced by trees; people linger and shop longer when trees are present.

49. Trees generate jobs and contribute raw materials for buildings, newspapers and books.

50. Wood by-products are used in a number of products such as plastics, vitamins, toothpaste and medicine.

Threats to urban forests

Despite their importance, urban forests face considerable threats.

- Budget constraints often limit ability of municipalities to hire skilled staff and dedicate adequate resources to properly care for trees.
- Urban forests often have low **species diversity**, are not planted with species tolerant to climate changes, and are vulnerable to pests and disease.

- In some cities, many of the large trees responsible for the bulk of important ecosystem services are legacies from the pre-settlement landscape and these legacy trees are now reaching the end of their natural lifespans.

We will lose much of the valuable services trees provide if we do not move to improve our ability to manage and care for the urban forest.

Research is required to confront threats and improve urban forests

Our aim should be a diverse, resilient, and climate change-ready urban forest. To achieve this, we must understand how to grow trees successfully in these areas; determine how they function within the surrounding ecosystem; develop new tree varieties with traits for urban conditions; accurately quantify the value of trees; devise improved planting and care techniques for urbanized regions; and, optimize urban forests for an uncertain future climate.

Protecting Global Tree Diversity

Worldwide, trees in the wild have tremendous ecological, economic, and cultural importance. They are the backbones of terrestrial biodiversity and provide vital ecosystem services like soil stabilization, groundwater purification, and nutrient cycling. Trees represent a rich evolutionary history and genetic reservoir, as roughly one in four flowering plant species is a tree. Globally, forests store more than 650 billion tons of carbon, more carbon than is present in the entire atmosphere. Trees are also valuable sources of fuel, fiber, medicine, timber, and food for people.

Threats to tree species in the wild

Globally, 8,000 species, or 10 percent of all tree species, are threatened or endangered related to many factors such as:

- Habitat loss,
- Overexploitation,

- Pressure from invasive species and,
- Climate change.

Protecting global tree diversity requires an approach that incorporates a wide variety of complementary techniques. These conservation strategies must be informed by tree science-based research including:

- Species distribution mapping,
- Reproductive biology,
- Phylogenetics,
- Germplasm storage,
- Micro propagation and tissue culture,
- Population genetics,
- Conservation collection assessment,
- Reintroductions and,
- Restoration ecology.

Public gardens can leverage their expertise and engage in larger research collaborations

Public gardens and arboreta have an increasingly important role to play in global tree conservation, yet their resources often are underutilized toward these ends. The Center for Tree Science facilitates tree conservation by supporting targeted research efforts and catalyzing global collaborations that combine diverse perspectives, expertise, and resources to safeguard tree species from extinction.

Considering the benefits trees provide, the threats they face, and the creative and challenging research needed to confront and overcome these threats, tree science requires increased support. Yet, in fact, support is declining. Resources for tree science research and training have been decreasing for two decades, diminishing the quantity of research and the number of students pursuing careers in tree science.

Chapter 9

Becoming a Tree Scientist

Tree scientists, called foresters (Figures 41a and 41b), plan, maintain, and preserve forests and forest resources for public and private use. They may protect wilderness areas, enhance habitats for a variety of species, facilitate public recreation, and manage fires. Many foresters help manage public areas, balancing wilderness protection and recreation in beautiful, inspiring, natural areas with timber needs.

Figure 41b Forester at work

Figure 41a Forester at work

Foresters may work closely with professionals in other fields, such as entomology, geology, hydrology, soil science, and wildlife biology. Foresters share ideas and develop plans with other agencies, as well as with citizens' action groups.

Professional foresters may multitask across several areas of responsibility, or specialize in a particular area such as soil science, fire suppression, or pest management. Foresters typically develop or manage plans for:

- Wilderness protection,
- Timberland improvement,
- Habitat enhancement,

- Timber sales,
- Tree nursery operations,
- Public recreation and,
- Fire management.

Forestry technicians help implement these measures by:

- Working on a fire suppression team,
- Constructing or maintaining trails,
- Maintaining campground facilities and assisting visitors,
- Monitoring forest use and enforcing regulations,
- Planting trees and shrubs to stabilize soils,
- Conducting inventories, classifying species and resources, and assessing forest health and,
- Marking trees to be cut, or preserved.

Foresters almost always work outdoors improving trails, assessing forest health and resources, or involved in a variety of other activities. Researchers and teachers may work in labs and classrooms.

The amount of human interaction can vary, depending on duties. While some foresters work primarily in backcountry areas, others who work at campgrounds, in education, or fire suppression interact with others on a daily basis.

With over 30,000 permanent employees in hundreds of locations across the country, the USDA Forest Service is a major employer of foresters. Many

Forest Service employees work in forest and range research, and others advise state and private partners across the country. Foresters also work at the state government level.

Foresters may also work in private industry for forest consultancies, timberland management companies, forestry firms and lumber companies. They may also work for conservation organizations, or teach at colleges and universities.

The West coast (California, Oregon, Washington, and Idaho), the Great Lakes area (Minnesota, Wisconsin), the Southeast (Florida, Alabama, North Carolina, Virginia), and Pennsylvania are the areas employing the largest numbers of foresters.

Forester careers involve all aspects of the forest lifecycle including harvesting, restoration, protection of fragile areas, and recreation. Duties vary from job to job, but the following list includes typical job duties one might encounter as a forester:

- Facilitate reforestation plans from inception to completion and management,
- Assist in selective cutting or clearing operations,
- Assess forests for reclamation and rehabilitation work,
- Create biomass resources like nurseries, sites, seeding, planting, controlling weeds, and pruning,
- Harvest seeds and cultivate young trees,
- Review research and literature relating to current discoveries and best practices,
- Review health and safety data for the forest, region, and time of year,
- Collect field and control samples of ecology and non-living media for analysis,

- Measure forest metrics on an ongoing basis,
- Enforce fire regulations,
- Survey and map forest areas and access roads,
- Encourage new growth,
- Use herbicides, insecticides and silvicides to implement pest, weed, and disease control,
- Monitor forest operations from a regulatory compliance perspective,
- Issue environmental field reports,
- Work in remote locations in all weather conditions,
- Process forest specimens and samples and,
- Have working knowledge of all tree species and their characteristics, as well as their ecological impact on the forest

Senior forester careers focus on leading teams with their management and administrative skills. Some of the additional tasks that accompany this senior position are:

- Approve the use of herbicides, silvicides and insecticides for use in the forest,
- Oversee sample management,
- Ensure accuracy in data recording,
- Use computer modelling to analyze forest data and predict outcomes,
- Collaborate with other industry professionals to contribute information to long-term management plans and reporting procedures,

- Communicate with clients, government departments, and the public,
- Oversee clearing actions and replantation actions,
- Oversee quality and safety controls for all timber handling processes including sawmilling, storage, transportation procedures, and inspection, sorting, and drying and,
- Have working knowledge of all species and their impact on local ecology.

Professional forestry positions generally require a bachelor's degree. Forestry programs include foundational courses on biology, chemistry, and other physical sciences, as well as courses on natural resources and environmental sciences. Since most professionals use GIS/GPS and remote sensing skills, courses on these technologies are also beneficial. Many universities and colleges offer baccalaureate degrees in forestry accredited by the Society of American Foresters (SAF). College graduates may be hired by the USDA Forest Service at the GS-5 or GS-7 grade level. After a year or two in training and developmental positions, they may be promoted to the GS-9 grade level. Master's or Ph.D. degrees are required for research positions.

Many forest technician positions require a two-year degree from a technical college recognized by SAF. However, many technical positions within the Forest Service don't require college degrees, while still offering opportunities for advancement. Foresters and forest technicians may need to be licensed or registered in some states. Voluntary professional certification is available through SAF.

Misty Pine Forest

Glossary

Apical – the tip, point, or vertex; summit.

Breast Height –A standard height from ground level for recording diameter, girth, or basal area of a tree, generally 4.5 feet.

Cone – the more or less conical multiple fruit of the pine, fir, etc., consisting of overlapping or valvate scales bearing naked ovules or seeds.

Ecosystem – an ecosystem is made up of plants, animals, microorganisms, soil, rocks, minerals, water sources and the local atmosphere interacting with one another.

Erosion – the act in which earth is worn away, often by water, wind, or ice.

Germplasm – living tissue from which new plants can be grown.

Meristems – are responsible for the division of new cells... they are zones of actively dividing cells.

Phloem – the living tissue that transports the soluble organic compounds made during photosynthesis.

Photosynthesis – the process by which plants and some bacteria use the energy from sunlight to produce sugar, which cellular respiration converts into ATP, a small molecule used in cells to transports chemical energy within cells for metabolism, the "fuel" used by all living things. The conversion of unusable sunlight energy into usable chemical energy is associated with the actions of the green pigment chlorophyll. Most of the time, the photosynthetic process uses water and releases the oxygen that all living things absolutely must have to stay alive.

Phylogenetics – the study of evolutionary relationships among biological entities - often species, individuals or genes.

Pole Stage – is a tree that is 3–12 inches in diameter **breast height**.

Sapling Stage – A tree, usually young, that is larger than a seedling but smaller than a pole. Size varies by region.

Silvicides – special pesticides used to kill brush and trees.

Species diversity – the number of different species that are represented in a given community.

Tree – a perennial plant with an elongated stem, or trunk, supporting branches and leaves or needles in most species.

Vascular cambium – a plant tissue located between the xylem and the phloem in the stems and roots of vascular plants. It is a cylinder of unspecialized meristem cells that divide to form secondary vascular tissues. It is the source of both secondary xylem growth inwards towards the pith, and secondary phloem growth outwards to the bark. Unlike the xylem and phloem, it does not transport water, minerals or food through the plant.

Valvate scales – bud scales can be arranged either in pairs, facing each other edgewise (valvate bud scales) or overlapping like shingles (imbricate scales). A few species have naked buds that lack scales and are protected instead by a pair of miniature leaves.

Xylem – one of the two types of transport tissue in vascular plants, phloem being the other. The basic function of xylem is to transport water from roots to shoots and leaves, but it also transports some nutrients.

Pine Forest

Sources

Dwyer, J.F., E.G. Mcpherson, H.W. Schroeder, and R.A. Rowntree. 1992. Assessing the benefits and costs of the urban forest. Journal of Arboriculture 18: 227–234.

Elmendorf, W., T. Watson, and S. Lilly. 2005. Arboriculture and urban forestry education in the United States: Results of an educators survey. Journal of Arboriculture 31: 138–149.

Fowles, John. 1979. The tree. Harper Collins, New York, NY.

Hayes, V. 2013. Four case studies: Public gardens fill the void in botanical education. Public Garden 28: 6–9.

Kalman, Bobbie and Kathryn Smithyman. 2002. The life cycle of the tree. Crabtree Publishing Co., New York, NY.

Keathley, D.E. 2003. National graduate education needs and priorities in natural resources. An assessment by the National Association of Professional Forestry Schools and Colleges, October 2003. National Association of University Forest Resources Programs.

Paxton, Jim. 2015. Forests, fires and conservation. Arizona Wildlife Views, May-June, 2015:18-24.

Ridsdale, Colin, John White, and Carol Usher. 2010. Visual reference guides:

Trees. Metro Books, New York, NY.

Suzuki, David and Wayne Grady. 2004. Tree, a life story. Greystone Books, Vancouver, British Columbia. Canada.

Trees of Christmas. 1988. National Wildlife Federation, Washington, DC.

Tudge, Colin. 2006. The tree, a natural history of what trees are, how they live and why they matter. Crown Publishers, New York, NY.

Wells, Diana. 2010. Lives of trees; an uncommon history. Algonquin Books of Chapel Hill, Chapel Hill, NC.

http://www.mortonarb.org/science-conservation/center-tree-science-securing-future-trees#improving

http://idahoptv.org/sciencetrek/topics/trees/facts.cfm

http://www.kidzone.ws/plants/trees.htm

http://www.environmentalscience.org/career/forester

http://www.care2.com/causes/50-benefits-of-trees.html

Snowy Pine Forest

Illustration Credits

https://commons.wikimedia.org/wiki/File:Pinus_hwangshanensis_trees.jpg

http://www.charlotteobserver.com/news/local/article16507784.html
(RALEIGH) NEWS & OBSERVER FILE PHOTO

www.naturetracking.com

http://www.bentler.us/eastern-washington/plants/trees/ponderosa.aspx

azdailysun.com TOM DeGOMEZ Special to the Daily Sun

<!-- HTML Credit Code for Can Stock Photo -->

<a href="http://www.canstockphoto.com">(c) Can Stock Photo / skylight</a>

http://waynesword.palomar.edu/ww0604.htm

https://www.fs.usda.gov/

https://www.thespruce.com/woodworking-with-pine-3536911

http://www.nhptv.org/natureworks/pileatedwoodpecker.htm

http://www.statesymbolsusa.org

http://www.arkive.org

http://www.amazingwallpaperz.com

7te.org

Pine branch

About the Author

Mary Jo Nickum is a retired science librarian. She has a B. A. in English education, a master's in librarianship and an interdisciplinary master's in fisheries science, anthropology and education. Her first book for children was a chapterbook, Mom's Story, a Child Learns about MS. She specializes in writing science for children.

Mary Jo lives in Fountain Hills, Arizona with her husband. She has two grown sons.

Made in the USA
Monee, IL
05 November 2026

67387145R00055